Tundra Animals

By Christopher Butz

STECK-VAUGHN
A Harcourt Company

www.steck-vaughn.com

Published by Raintree Steck-Vaughn Publishers, an imprint of Steck-Vaughn Company.

Library of Congress Cataloging-in-Publication Data
ISBN: 0-7398-6410-6

Printed and bound in the United States of America
1 2 3 4 5 6 7 8 9 10 WZ 06 05 04 03 02

Produced by Compass Books

Photo Acknowledgments
Corbis, cover, 2, 4, 8, 16, 20, 22, 45; Digital Stock, 11, 13, 28, 44; Fritz Pölking, 14; Richard Walters, 18; Anthony Mercieca, 24; Tom Walker, 27: Hugh Rose, 30; Richard Herrman, 32; Brandon Cole, 35; Carolina Biological, 39, 45; S. McCutheon, 42; Jeff Foott, 40; Thomas Kitchin, 36.

Content Consultant
Dr. Harry N. Coulombe
USGS–Patuxent Wildlife Research Center
Laurel, MD

This book supports the National Science Standards.

Contents

Animals on the Tundra. .5

The Emperor Penguin. 9

The Mosquito. 17

The Caribou. 25

The Sockeye Salmon. 33

What Will Happen to Tundra Animals?. 41

Quick Facts. 44

Glossary. 46

Addresses, Internet Sites, Books to Read. 47

Index. 48

This arctic fox is an animal that lives in the cold tundra biome.

Animals on the Tundra

The Arctic tundra is a treeless area between the polar icecap and the treeline to the south. Most people think that snow and ice are the only things on the tundra. But it is a **biome** and a home for many living things. A biome is a large region, or area, made up of communities. A **community** is a group of certain plants and animals that live in the same place.

The tundra is very cold. It does not receive much rain or snowfall during the year. There is a lot of wind on the tundra and no trees to block it.

Arctic tundra is found in northern Europe, northern North America, the coast of Antarctica, and on some islands in the Arctic Ocean. Alpine tundra occurs in patches on mountaintops.

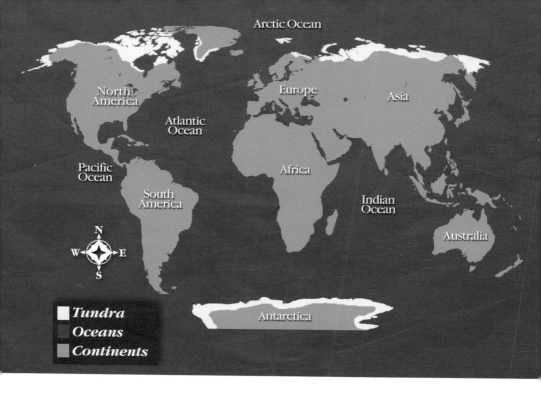

Arctic Ocean

North America

Europe

Asia

Atlantic Ocean

Pacific Ocean

Africa

South America

Indian Ocean

Australia

N
W — E
S

■ *Tundra*
■ *Oceans*
■ *Continents*

Antarctica

One-fifth of Earth's surface is tundra. This map shows where tundra is located around the world.

What Lives on the Tundra?

Living things have adapted to the tundra. To **adapt** means that a living thing changes over time in order to fit well in the conditions of the place it lives. The tundra's long winters can last for ten months. Its short summers are just warm enough for the tundra's special animals and plants to grow.

Unlike other biomes, tundra has permafrost. Permafrost is a layer of frozen ground. It begins just below the surface and can reach more than 1,000 feet (300 m) deep. The tundra's cold climate keeps this layer of ground frozen all year round.

Most plants cannot survive the tundra's cold temperatures. Plants on the tundra grow quickly during the short warm season. Then the plants must live through the long cold season or leave seeds to grow new plants.

There are fewer kinds of animals on the tundra than in the other biomes. Most of the animals **migrate**. To migrate means to move to a warmer area when winter approaches and then to come back in the summer.

In the next chapters, you will learn about four kinds of tundra animals. Emperor penguins are birds that spend much of their lives in the ocean. Mosquitoes feed on the animals of the tundra during the short summer. Caribou migrate thousands of miles every year. Sockeye salmon live in the ocean for years, then return to the rivers where they hatched. Keep reading to find out how each of these animals has adapted to live in its tundra home.

You can tell that this emperor penguin is old because the ring around its neck is orange.

The Emperor Penguin

Penguins are birds. Unlike most other birds, they cannot fly. But they can swim faster than most animals that would try to eat them.

Emperor penguins are the largest kind of living penguin. Adults weigh from 42 to 101 pounds (19 to 46 kg). They have large heads, short necks, and small wings. Skin connects the penguins' toes. These webbed feet help them swim.

The emperor penguin has special markings. Its dark beak has a streak of orange on the bottom. The front of the emperor penguin is white, and its back is bluish gray. Its head is black, but it has a patch of bright yellow feathers next to its ears. It also has a yellow circle around its neck that becomes orange as the penguin grows older.

Where Do Emperor Penguins Live?

Most emperor penguins live on the tundra of the southern hemisphere. The equator is an imaginary line that wraps around Earth, separating the northern half from the southern half. Each half is called a hemisphere.

For part of the year, emperor penguins live on the edge of the tundra by or on the ice packs. Ice packs are thick masses of sea ice that are like islands. Ice packs form along the ocean coast over the water.

Emperor penguins are social animals. This means they live together in groups. A group of emperor penguins is called a colony. The size of the colony ranges from 200 pairs of penguins to 50,000 pairs. Members of the colony live together, hunt together, and travel together. They often swim in the ocean to hunt for food.

How Have Emperor Penguins Adapted to Live on the Tundra?

Emperor penguins are specially adapted to live on the tundra ice packs. A thick layer of fat under their skin, called blubber, holds in body heat and keeps them warm.

This group of emperor penguins lives together on the tundra.

The emperor penguin's feathers also help keep it warm. It has two kinds of feathers. The outer feathers are shiny and waterproof. The inner feathers are down. Down is soft, warm, and fluffy. The outer feathers cover and keep the down dry.

Swimming

Emperor penguins are suited to swim in the oceans. Oceans are saltwater, which means they contain salt. To get the extra salt out of their bodies, penguins have a pair of glands behind the beak. A gland is an organ, or body part, that releases something. Emperor penguins' glands release extra salt. Because of these glands, the penguins can drink saltwater safely. Humans, and many other birds, would die if they drank only salty ocean water.

Penguins are too heavy to fly through the air. However, they are such good swimmers that some people say they fly through the water. To move through the water, a penguin uses its wings the same way other birds use their wings to fly.

What an Emperor Penguin Eats

Emperor penguins are carnivores. A **carnivore** is an animal that eats only meat. Seafood makes up all of an emperor penguin's diet. Common foods are fish, squid, and shrimp.

Penguins must hunt for their food in the water. Like all birds, penguins breathe air. They cannot stay underwater for a long time or dive

These emperor penguins are swimming underwater to look for food.

too far underwater. Penguins do not usually stay underwater for more than three minutes at a time. They swim to the surface when they need to breathe again. Then, they dive back underwater to continue to hunt for food. Emperor penguins dive only to a depth of 65 or 70 feet (about 20 m).

▲ This emperor penguin is feeding its chick.

An Emperor Penguin's Life Cycle

Emperor penguins mate in late April or early May. A female lays only one egg each year, usually in May. Then the female swims out to sea to hunt. During this time away, she eats a lot of food.

The male emperor penguin stays to incubate the egg. To incubate is to keep the egg warm.

The emperor penguin cannot set the egg on the ice or cold ground because the egg would freeze. To incubate the egg, the male places it on his feet. He covers the egg with his brood patch to keep it warm. The brood patch is a thick layer of blubber covered with warm feathers. Because he cannot leave the egg, he does not eat for about two months while he incubates the egg.

The chick hatches from the egg sometime in July. At this time, the mother returns. To feed the newly hatched chick, she regurgitates some of the food she has eaten. To regurgitate is to bring the food back up out of her stomach into her mouth. She feeds this food to the chick.

Both the male and female take care of the chick. The young penguin does not have the slick adult feathers that keep it dry underwater. It cannot hunt fish for itself. The parents must leave the chicks to hunt for fish. When they do, they leave their chicks with a group of other chicks. This group is called a crèche (KRESH).

Over the next few months, the young penguins grow their adult feathers. By the middle of summer, they are large enough to survive on their own.

You can see this mosquito's head, thorax, and abdomen.
It is resting near a dewdrop on a blade of grass.

The Mosquito

The mosquito is one of the most common insects in the world. An insect has three pairs of legs and three main body sections. These are the head, thorax, and abdomen. Most insects also have wings and feelers called antennae.

Mosquitoes are small. Most are normally less than one-half inch (1.5 cm) long. They have a long, thin body and long wings.

There are about 3,500 different kinds of mosquito. Fewer than 12 kinds of mosquito live on the tundra. During summer, there are more mosquitoes on the tundra than anywhere else in the world. Clouds of mosquitoes can be so thick that they seem to turn the sky black.

> This female mosquito is drinking blood so that she can reproduce.

Where Do Mosquitoes Live?

Mosquitoes live in many places around the world. Tundra mosquitoes live throughout the tundra. They begin life in still water.

The tundra has many pools of still water. The summer on the tundra is short, lasting only six to ten weeks. During this time, the snow and ice melt. Because of the frozen permafrost, the water cannot soak into the ground as it would in a warmer climate. This causes great pools of still water to form on the tundra.

How Have Mosquitoes Adapted to Live on the Tundra?

Mosquitoes that live on the tundra have adapted to the cold, dry winters. They grow very quickly during the short summer. Then they lay their eggs before winter arrives. The cold winter would kill the eggs of most animals. But the eggs of tundra mosquitoes lie dormant during winter. Dormant means inactive. The eggs become active again the following summer when they are surrounded by water. At that time, the eggs hatch.

The mosquito also needs warm blood to reproduce. During the summer, larger herds of animals, such as caribou and musk ox, travel to the tundra. Female mosquitoes drink the blood of these animals, as well as birds, rabbits, and lemmings. Lemmings are small mouse-like animals that look like hamsters.

You can see this mosquito's proboscis. This close-up picture was taken with a microscope.

What a Mosquito Eats

Just like other mosquitoes all over the world, mosquitoes of the tundra feed on nectar. Nectar is the sweet, sugary liquid found in flowers. Thousands of flowers bloom during the short tundra summers.

Female tundra mosquitoes also drink blood from warm-blooded animals. Warm-blooded animals have a body temperature that usually stays the same, no matter what the temperature is outside. Although female mosquitoes drink blood, it is not for food. They need blood only to produce eggs.

The mosquito has a long tube attached to its mouth. This tube is called a proboscis. The tube is used to suck liquids into the mosquito's mouth.

The proboscis of the female mosquito is different from the male's. It is sharper and allows her to stab through an animal's skin and drink its blood. Only female mosquitoes drink blood.

FUN FACT

Sometimes mosquito attacks make herds of caribou stampede. A stampede is a sudden rush of scared animals. The caribou will stampede to try to escape the thick clouds of female mosquitoes.

▲ These mosquito larvae are floating upside down in the pond where they hatched.

The Life Cycle of the Mosquito

Adult mosquitoes mate during the tundra's short warm season. After mating, the females bite animals. The warm blood allows them to develop eggs. Then the female lays eggs that will hatch the following summer.

The mosquito goes through four growth stages. The first stage is as an egg. When the weather warms, the egg hatches. The mosquito then passes through the second stage of life as a larva.

Mosquito larvae are long and legless. They have a head at one end and an opening for breathing at the other end. The larvae float upside down in water with the breathing end poking out of the water's surface. The head is underneath the water, and the larvae eat tiny pieces of floating plants.

The mosquito larva molts four times. Molting is when an animal sheds its skin. A mosquito has an exoskeleton. This hard outer covering supports its body. An insect does not have a bony skeleton inside its body like a person does. Since exoskeletons do not grow, the insect must shed the exoskeleton in order to grow larger.

After the fourth molt, the mosquito is in the third stage of its life cycle. It is now a pupa. Mosquito pupae remain floating in the water. They breathe through new tubes in the upper back. After a few days, the pupa is fully formed. The mosquito molts once more and steps onto the surface of the water as an adult.

The set of antlers on a fully grown caribou is called a rack.

The Caribou

The caribou is a **mammal** that lives on the tundra. A mammal is a warm-blooded animal with a backbone. Female mammals give birth to live young and feed them with milk from their bodies.

Caribou are large mammals. They are up to 5 feet (1.5 m) tall and weigh up to 700 pounds (318 kg). Caribou have long, thin muscular legs with thick hooves to protect their feet. Their necks are long and thin. They have a long, shaggy tan or brown coat with white patches on the face, chest, and backside. Both males and females grow antlers. A set of antlers may reach up to 4 feet (1.2 m) wide.

Where Do Caribou Live?

Caribou live on the tundra in Alaska, northern Canada, Greenland, Scandinavia, Siberia, and Baffin Island. In Scandinavia and Siberia, caribou are called reindeer.

The caribou of North America are still wild. Like most animals of the tundra, the caribou migrate. During the winter, the caribou travel hundreds of miles south. There, they live in forests. Food is easier to find in forests than on the tundra during the cold months of winter.

When summer comes, the caribou start their long migration north back to the tundra. On the tundra are the calving grounds. A calving ground is an area where the caribou cows go to give birth to their calves. In this case, a cow is a female caribou. In the autumn, the caribou will begin to migrate back to the forests for the winter.

How Have Caribou Adapted to Live on the Tundra?

Caribou are suited to traveling over the tundra. Their wide hooves act like snowshoes to help keep them from sinking into the snow.

▲ This herd of caribou is crossing a lake during their migration.

When traveling through snow, caribou follow in each other's hoofprints. This helps the caribou save energy, so they do not need as much food.

Caribou have ways to stay warm on the tundra. A thick coat of fur covers their bodies. The hairs trap air, which helps hold in their body heat.

This caribou has found grassy plants to eat during the tundra's short summer.

What Do Caribou Eat?

Caribou are herbivores. Herbivores eat only plants. Caribou will eat whatever food they can find growing on the tundra.

Summer is the main feeding season for caribou. This is when food is easy to find.

During the summer, many kinds of plants grow. These include grasses and mushrooms. The caribou must eat enough food to store fat for the winter. The caribou uses energy from stored fat when it cannot find food to eat.

On the tundra, there is snow or frost on the ground most of the rest of the year. During this time, the main food for caribou is lichen. This crust-like plant is a combination of moss and fungus that grows on stones. To find lichen, the caribou brushes the snow away with its hoof. Then, it scrapes the lichen off the stone and eats it.

Caribou are ruminants. A ruminant has four stomachs and a different way of digesting food than other mammals. To **digest** is to break down food so the body can use it. When the caribou eats, food enters its first stomach, where digestion begins. Then the caribou regurgitates its food and chews it again for several minutes. This is called chewing its cud.

After the caribou is finished chewing its cud, it swallows the chewed-up ball of partly digested food. This goes into its second stomach. From there, it passes into the third and fourth stomachs, which finish the digestion.

▲ These young caribou can walk and travel with the herd. Their antlers grow as they get older.

The Life Cycle of the Caribou

Caribou live together in herds. Herds of caribou can be very large. Some have thousands of animals.

During the summer, both male and female caribou begin to grow antlers. Antlers are bones that come out of the skull. A soft skin called velvet covers growing antlers. By autumn, the antlers reach their full size. When this happens, the velvet dies. The caribou scrapes the dead velvet off its antlers by rubbing them against trees and rocks. Once this happens, it is time for mating season. After mating season, the antlers fall off.

Mating season usually begins in October. As it approaches, bulls (male caribou) begin fighting for the right to mate with the cows (female caribou). When the bulls fight, they use their antlers to try and drive the others back. Bulls usually are not hurt in these fights. The winning bulls mate with the cows.

In April, the caribou begin their migration north. Only the cows about to give birth travel to the calving grounds.

Females give birth to one or two calves usually in May or June. The calves are able to stand within minutes of birth. They must stand in order to drink milk from their mothers. Within two hours, they can walk. They must walk to catch up with the rest of the migrating herd.

You can tell that these sockeye salmon are old because of their red back and sides.

The Sockeye Salmon

A sockeye salmon is a fish. A fish is a cold-blooded animal with a backbone. Cold-blooded animals have a body temperature that changes, depending on the temperature of the air or water around them. A fish's body is shaped so it can swim easily through water.

Sockeye salmon are different sizes depending on the place they hatched. They are from 3.5 pounds (1.6 kg) to 7 pounds (3.2 kg) in weight.

The adult sockeye salmon has a greenish-blue back with small black spots. Its sides are bright silver. Because of its coloring, the sockeye salmon is also called the blueback salmon. As it gets older, the color changes. Its head then turns olive green, and its back and sides turn dark red.

Where Do Sockeye Salmon Live?

Sockeye salmon start their lives in freshwater rivers or lakes of the Pacific Northwest, Alaska, northwestern Canada, and southeastern Siberia. Some sockeyes will never leave the lakes where they grew up.

Sockeyes that hatch in rivers and other lakes swim out into the Pacific Ocean for several years. They migrate around the Northern Pacific in the Sea of Okhotsk, Anadyr Gulf, Bering Sea, Bristol Bay, and the Gulf of Alaska.

After spending one or more years at sea, the sockeyes return to the rivers or lakes where they hatched in order to **spawn**. Spawn is to lay large numbers of eggs to produce young.

How Have Sockeye Salmon Adapted to Live in the Tundra?

Sockeye salmon are suited to live in the rivers and lakes of the tundra. Like all animals, sockeyes need oxygen to live. But they do not breathe air in the same way that people do. Instead, they use body parts called gills to absorb, or take in, oxygen from water.

Scales protect the sockeye salmon from the cold water.

A sockeye salmon's body is made to live in the water. It has side fins that help it change direction in the water. It pushes itself forward by moving its tail fin back and forth.

Sockeye salmon are covered with scales. A scale is a small, plate-like piece of thick skin. Scales help hold in heat and keep the fish warm.

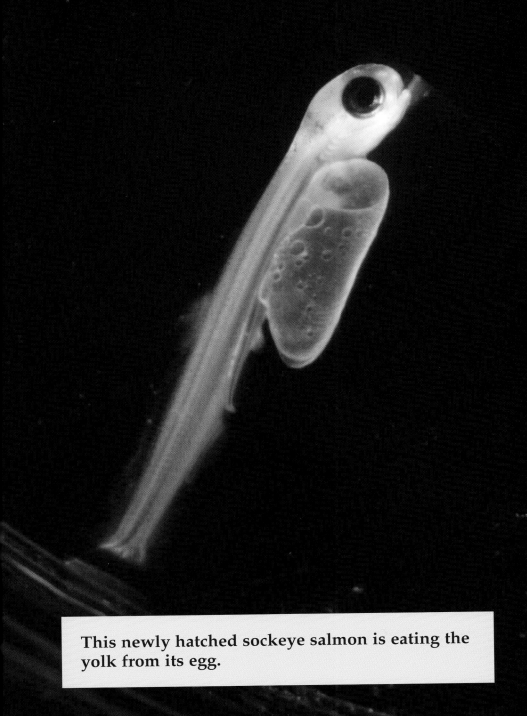

This newly hatched sockeye salmon is eating the yolk from its egg.

What a Sockeye Salmon Eats

Sockeye salmon eat different things depending on the stage of their life cycle. When they are newly hatched, they eat the yolk from their egg.

As they grow, the young sockeye salmon are omnivores. An omnivore eats both plants and animals. Sockeye salmon eat plankton. Plankton are small plants that float in water. They also eat small insects.

As adults, sockeye salmon are carnivores. Common foods are snails, crab larvae, and zooplankton. Zooplankton are microscopic animals that float in water. Microscopic means very small. They also may eat fish that are smaller than they are.

Sockeye salmon are an important part of the tundra's food chain. A food chain is an ordered arrangement of animals and plants in which each feeds on the one below it in the chain. Sockeye salmon eat smaller fish, but larger fish eat sockeye salmon. Even mammals, such as bears, catch sockeye salmon to eat when the fish swim up the rivers to spawn.

A Sockeye Salmon's Life Cycle

In late summer or autumn, sockeye salmon spawn. To do this, males and females swim from the ocean back to the river or lake where they first hatched.

Female sockeye salmon use their tails to dig nests in the gravel beds of rivers or lakeshores. The female lays up to 3,000 bright pink eggs in the nest. Then the male fertilizes the eggs. After this, the female covers the nest with gravel. After they spawn, the sockeye salmon live for only another week or two and then die.

When the eggs hatch, the tiny fish are called alevin. The alevin stay in the nest for three to four months.

The alevin leave the nest as small salmon called fry. Fry are about 1 inch (2.5 cm) long. They have a shiny green back, silver sides, and a white belly. Most fry swim to a lake where they will spend about one year before heading out to sea.

After a year, the sockeye salmon are called fingerlings. They are about 4 inches (10 cm)

> **During spawning, each female lays thousands of these colorful red eggs.**

long. The fingerlings swim from the river and into the ocean.

Once out at sea, the sockeye salmon grow to adulthood. They may spend up to four years out at sea, moving in great circular migrations every year. Then they return to their home river or lake to spawn and begin the cycle again.

Oil drilling like this may hurt the tundra. If there are accidents, spilled oil might kill wildlife.

What Will Happen to Tundra Animals?

The tundra provides a special habitat for the animals that live there. A habitat is a place where an animal or plant usually lives. Many plants and animals that live on the tundra could not live in other biomes.

In some ways, people have left the tundra alone. Few people live on the tundra because of its cold climate. Unlike other biomes, people have not destroyed large parts of the tundra to build homes and cities.

Still, there is one way that people harm the tundra—oil drilling. An oil spill hurts large areas of wilderness. Even careful drilling changes the areas where animals live. Oil spills can be extremely harmful to wildlife.

People are watching these sockeye salmon spawn. Fishing laws help keep the fish safe.

How Are Tundra Animals Doing?

Emperor penguins and mosquitoes are tundra animals that are doing well. However, some animals of the tundra are **endangered**. Endangered means a type of animal may die out in the wild.

Some kinds of caribou are close to extinction. Extinction means there are no more of those animals living. There are very few woodland caribou left in the wild. The main problem is habitat loss. People are tearing down the forests where these caribou spend the winters.

Overhunting also puts some kinds of caribou in danger. This means hunters have killed more caribou each year than the animals can produce through mating. Today, there are laws against hunting endangered caribou. This will help save caribou for the future.

More than 60 million sockeye salmon are caught and sold every year by professional fishermen. There are laws that limit catches. These laws have helped to keep the sockeye salmon from becoming an endangered species.

People must work to save the tundra. Roads and buildings are needed to discover and pump out oil. But these activities can damage the permafrost. Over time, this will hurt the tundra plants and animals. It is important to do what is best both for the tundra and for people. That way, tundra animals will be safe in their homes for many years to come.

Quick Facts

When it is very cold, emperor penguins gather close together in large circles. The heat from each other's bodies helps to keep them warm. They take turns in different places of the circle. Penguins on the outside of the circle are colder because they block cold winds and snow. The middle of the circle is the warmest place.

Male and female caribou have different names. Male caribou are called bulls. Females are called cows.

Mosquitoes belong to an order of insects called Diptera. All the insects within this order are flies with two wings.

Mosquitoes' eyes are multifaceted. This means that the eyes have many flat surfaces on them. Each of these surfaces helps the eye see in a different direction.

About 270 million sockeye fingerlings leave Lake Iliamna in Alaska over a period of just a few days.

Glossary

adapt (uh-DAPT)—to change to fit well in a particular environment

biome (BYE-ohm)—large regions, or areas, in the world that have similar weather, soil, plants, and animals

carnivore (KAHR-nuh-vor)—an animal that eats only meat

community (kuhm-YOO-nih-tee)—different species of plants and animals living together in a habitat

digest (dye-JEST)—to break down food so the body can use it as energy

endangered (en-DAYN-jurd)—a plant or animal species that is in danger of dying out

mammal (MAM-uhl)—a warm-blooded animal that breathes air, has a backbone, and gives birth to live young

migrate (MYE-grate)—to move from place to place according to the change of seasons

spawn (SPAWN)—to produce a large number of eggs

Addresses and Internet Sites

Arctic National Wildlife Refuge
101 Twelfth Avenue
Room 236
Fairbanks, AK 99701

Canadian Wildlife Service Environment
Ottawa, Ontario
K1A 0H3
Canada

Alaska Department of Fish and Game
www.state.ak.us/adfg/
geninfo/kids/kids.htm

Antarctic Penguins
www.gdargaud.net/
Antarctica/Penguins.html

Tundra Animals
mbgnet.mobot.org/sets/
tundra/animals/index.htm

Books to Read

Baxter, John M. *Salmon.*
Stillwater, MN:
Voyageur Press, 2000.

Nelson, Julie. *Tundra.*
Austin, TX: Steck-
Vaughn, 2001.

Index

alpine tundra, 5
antlers, 25, 31

biome, 5, 41
blubber, 10, 15

caribou, 7, 19, 21,
    25–31, 43
carnivore, 12, 37
colony, 10

emperor penguins, 7,
    9–15, 42

herbivores, 28

ice pack, 10
incubate, 14, 15

lichen, 29

migrate, 7, 26, 31, 34, 39
mosquitoes, 7, 17–23

omnivores, 37

permafrost, 7, 19
plankton, 37
proboscis, 21

ruminants, 29

scales, 35
sockeye salmon, 7,
    33–35, 37–39, 43
spawn, 34, 38, 39